STUDIO COMPANION SERIES

# DRAFTING BASICS

BOOK TWO

# STUDIO COMPANION SERIES

|||||||||||||||||||||||||||||||||||||||||||||||||||||||||||||||||||||||||||||||||||||||||

DESIGN BASICS/BOOK ONE

DRAFTING BASICS/BOOK TWO

3D DESIGN BASICS/BOOK THREE

PRESENTATION BASICS/BOOK FOUR

|||||||||||||||||||||||||||||||||||||||||||||||||||||||||||||||||||||||||||||||||||||||||

STUDIO COMPANION SERIES

# DRAFTING BASICS

BOOK TWO

DONNA LYNNE FULLMER
IIDA, IDEC

KANSAS STATE UNIVERSITY
DEPARTMENT OF INTERIOR ARCHITECTURE
AND PRODUCT DESIGN

FAIRCHILD BOOKS
NEW YORK

Executive Director & General Manager:  Michael Schluter

Executive Editor:  Olga T. Kontzias

Assistant Acquisitions Editor:  Amanda Breccia

Senior Development Editor:  Joseph Miranda

Assistant Art Director:  Sarah Silberg

Production Director:  Ginger Hillman

Associate Production Editor:  Linda Feldman

Ancillaries Editor:  Amy Butler

Copy Editor:  NaNá V. Stoelzle

Associate Director of Sales:  Melanie Sankel

Cover Design:  Carly Grafstein

Text Design:  Kendrek Lyons

Page Layout:  Tom Helleberg

All drafting tools and equipment in photographs supplied courtesy of Alvin & Company

Library of Congress Catalog Card Number: 2011915779

ISBN: 978-1-60901-095-9

GST R 133004424

Printed in the United States of America

TP09

# CONTENTS

||||||||||||||||||||||||||||||||||||||||||||||||||||||||||||||||||||||||||||||||||||

# THREE – ORTHOGRAPHIC DRAWINGS

# FOUR – ARCHITECTURAL LETTERING

# PREFACE

The overall idea for this series of books came from the love of teaching the freshman design studio. What I have seen, time and time again, is a lack of basic design skills due to the influx of technology. While I believe in being a well-versed designer, I feel the computer is just another tool in the arsenal of what a student, and a professional, can bring to the table when it comes to designing. To this end, I feel hand skills, and the teaching of hand skills, has become a lost art.

I tell students that beginning the study of architecture and design is like starting kindergarten again, because we ask them to learn to write and draw in a new way. The books in the Studio Companion Series acknowledge that and act as an introduction to a skill through interactive lessons for each topic. I have seen firsthand how students increase their skills more rapidly by doing rather than just seeing. In addition, like it or not, today everyone wants things faster and easily digestible. I feel this format, with a lot of images and text that is direct and simple to read, will play to this audience of future designers.

The Studio Companion Series includes four books that address all the skill sets or topics discussed in beginning the study of architecture and design. Each book is compact and highly portable, and addresses each topic in a clear-cut and graphic manner. Each has been developed for today's students, who want information "down and dirty" and presented in an interactive way with simple examples on the topics. The series includes *Design Basics, Drafting Basics, 3D Design Basics,* and *Presentation Basics.*

Students have been drawing lines, writing their alphabets, and using a ruler for years now, and all of a sudden they are being asked to change the way they do it. Technical drafting is the formal way an architect and designer draws his or her intentions, and it is truly an art form. In *Drafting Basics*, students will learn to use the correct drafting tools properly, draw lines and two-dimensional drawings successfully, and letter architecturally.

This book is the result of years of teaching students all over the country and listening to their questions when they were using standard textbooks.

# ACKNOWLEDGMENTS

||||||||||||||||||||||||||||||||||||||||||||||||||||||||||||||||||||||||||||||||||||||||||||||||||||||||||||||||||||||||||||||||||

The Studio Companion Series represents the result of working
with students all over the country and the thrill I get watching the
"lightbulbs go on" as they learn. There is nothing like seeing a student
use a scale properly for the first time—it is an addiction to be a part of
this type of learning. To all my students, thank you for giving me that
charge and making me proud!

Two students, Jeff Snyder and Kyle Emme, specifically helped with this
book and deserve to be singled out. They are tremendous students,
designers, and friends whom I have had the pleasure of knowing since
their freshman studios respectively. Thanks to Jeff for assisting with
the design of the cabin seen throughout the series and most of the
drawings both electronic and by hand and to Kyle for his help with
the physical model. Your enthusiam and dedication is infectious and
appreciated.

To my high school English teacher and to the men and women I work alongside every day, thank you for showing me how to listen to students and respond respectfully while maintaining the authority in the classroom.

Finally, everyone writes and says this, but I truly owe my career to my supportive and loving family, who have taught me things you could never find in a textbook.

STUDIO COMPANION SERIES

# DRAFTING BASICS

BOOK TWO

# ONE

## DRAFTING TOOLS

### OBJECTIVES

You will be able to identify and successfully use:

- The basic tools for technical drafting
- Architectural scales for reading and producing drawings

NO DOUBT YOU HAVE PURCHASED A SET OF TOOLS, USUALLY REFERRED TO AS A KIT, OF THE BASIC SUPPLIES YOU WILL NEED IN YOUR FIRST YEAR OF STUDY OF ARCHITECTURE AND DESIGN. MOST OF THESE TOOLS ARE USED FOR TECHNICAL DRAFTING BUT SOME ARE ALSO USED FOR BUILDING MODELS. IN THIS CHAPTER WE DISCUSS THE USE OF THE TOOLS FOR DRAFTING AND THE THIRD BOOK IN THE STUDIO COMPANION SERIES WILL ADDRESS THE TOOLS USED FOR MODEL BUILDING.

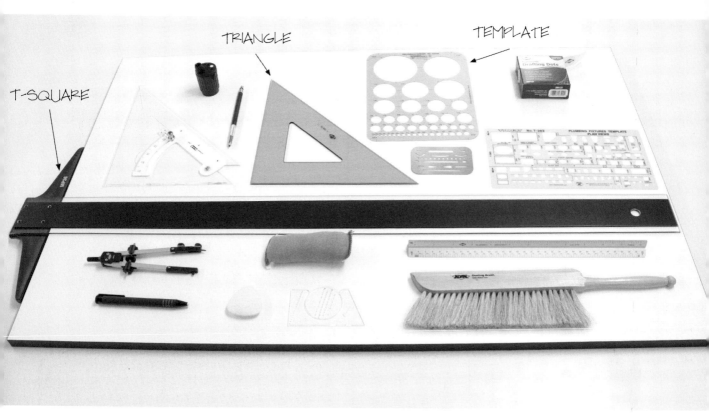

T-SQUARE

TRIANGLE

TEMPLATE

Typical drafting kit tools; each will be explained further in this chapter.

The tools used to draft are tried and true and will allow you to create all the drawings and letters needed to succeed. Learning the correct name and use of each tool is critical.

# T-SQUARE AND DRAFTING BOARD OR DESK

The first two tools are the most important and will be used daily. The first is a T-square, the large "T" shaped metal or wooden instrument. This is used in tandem with a drafting board or desk. The very first rule in drafting is to make sure your T-square is "locked and loaded" snugly to the left side of the drafting board. Not doing so will create inaccurate lines and angle relationships.

Other tools that can be used in place of a T-square include a drafting machine and parallel straightedges; both of which are permanently attached to a drafting table. Beginning design students typically start with a T-square to learn the proper use of the drafting tools.

Most of the other tools will rely on the T-square, or the two tools mentioned above, and the drafting board for their successful use. Focusing on the T-square as our tool of choice, we see that it is used to draw any horizontal lines while supporting the next three tools to create vertical and angled lines.

Drafting machine and parallel straightedges.

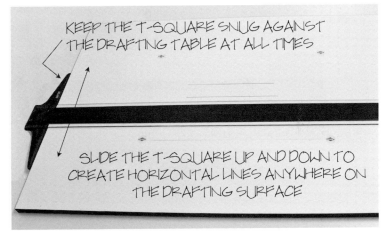

KEEP THE T-SQUARE SNUG AGAINST THE DRAFTING TABLE AT ALL TIMES

SLIDE THE T-SQUARE UP AND DOWN TO CREATE HORIZONTAL LINES ANYWHERE ON THE DRAFTING SURFACE

The T-square "locked and loaded" on the left side of the drafting board.

# TRIANGLES

||||||||||||||||||||||||||||||||||||||||||||||||||||||||||||||||||||||||||||||||||||||||||||||||||||||||||||||||||||

There are three basic triangles: a 30/60/90–degree triangle, a 45-degree triangle, and an adjustable triangle. The first two triangles, when placed on the T-square, will create a line at the angle in their description. The adjustable triangle can create any angle needed that the other two cannot create. In all cases the triangle must rest on top of the T-square to achieve the desired angle.

Most triangles have a raised edge, known as an inking edge. This allows you to draft with ink while not allowing it to seep under the triangle and ruin your drawing. Inking edges are also found on other tools for the exact same reason. If the tool is available with or without an inking edge, I recommend the inking edge every time.

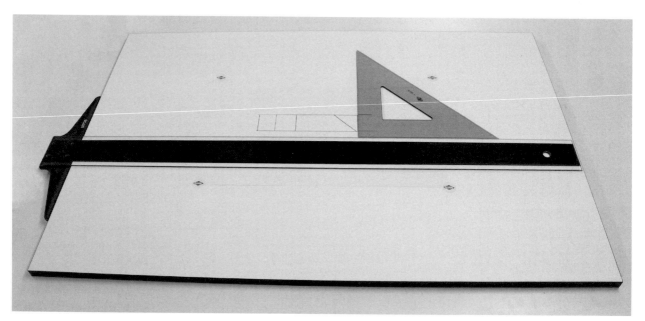

Triangles create vertical and angled lines.

# RULERS AND SCALES

||||||||||||||||||||||||||||||||||||||||||||||||||||||||||||||||||||||||||||||||||||||||||||||||||||||||||||||||||||||||||||||

A cork-backed, traditional-looking ruler may be included in your kit. This is to be used for model building, not drafting, and is cork backed to help it not slip while cutting. This will be discussed in book three of the Studio Companion Series.

The triangular-shaped ruler is called a scale and you may have both an architectural scale and a metric scale or even an engineering scale.

The scale is crucial as you create drawings and understanding how to use one is very important. Obviously we cannot draw a building to its real size and dimension; it would never fit on a piece of paper, so we use scales to allow us to draw the space proportionally correct. Reading a scale may seem confusing, at first, but once you know how to read one you will never forget.

A metric scale is used just like an architectural scale except for the metric system. An engineering scale is used on much larger drawings and will be used by consultants including all types of engineers, landscapers, planners, and architects. As a beginning design student we will focus on the architectural scale.

# UNDERSTANDING THE SCALE

As you look at each face of the scale you will notice several things. First, at each end you will see a number or fraction; this is the actual name of the scale.

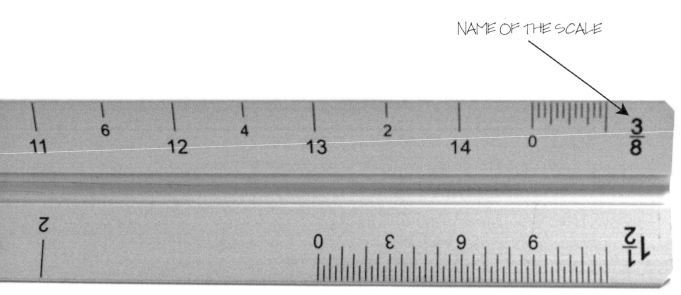

NAME OF THE SCALE

The number, a fraction in this example, on the right side of the zero is the name of the scale, in this case ⅜".

The name of the scale is also the increment that will be referred to as a foot when drafting. So, $3/8$" scale means that each $3/8$" of an inch is equal to a foot in a drawing. From this point on you should add a label to all drawings that indicates the scale in which it was drawn. It is typically written as follows and is placed in the lower left- or right-hand side and below the drawing.

Scale: $3/8$" = 1'-0"

Graphically, an apostrophe indicates feet while a quotation mark indicates inches.

# INCHES VERSUS FEET

The second thing you may notice is next to every scale name there is a series of small lines followed by a zero. This indicates the division of that scale into inches. Rotate the scale around and you will see that the ³/₄" scale has been divided into 24 small increments. This means each mark is equal to a half an inch—at that scale. Rotate the scale again to find the ¹/₈" scale and you will see it is only divided into six increments. This means each little line is equal to two inches.

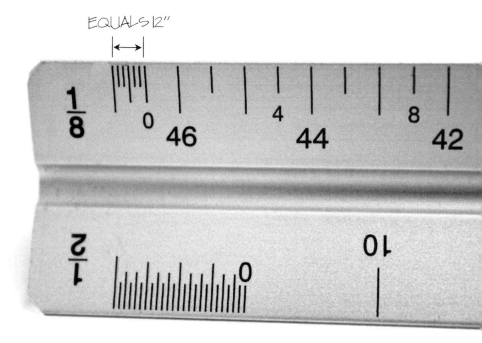

At ¹/₈" = 1'-0" each mark represents 2 inches.

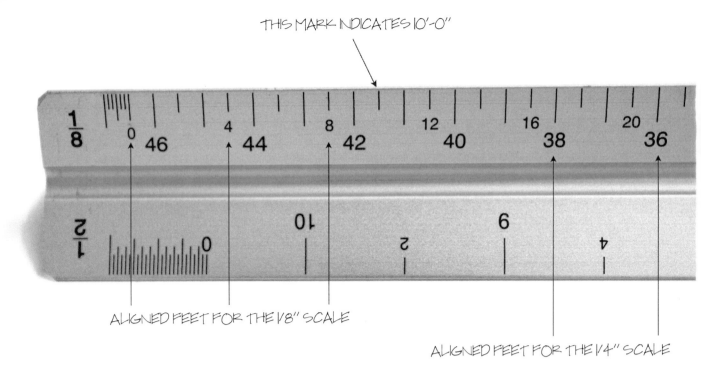

THIS MARK INDICATES 10'-0"

ALIGNED FEET FOR THE 1/8" SCALE

ALIGNED FEET FOR THE 1/4" SCALE

The location of the number for the feet measurement of the scale is aligned with the zero for that scale.

Third, you will see a series of numbers running all the way down the scale. These numbers are the feet for that scale. You will notice that all the numbers are located under a vertical line and that line is the same height as the line above the zero for that scale. This graphically makes it easier to follow which number belongs to which scale.

Look at the 3/4" scale and find the 10'-0" mark. This is a little difficult since the 10'-0" mark for the 3/8" scale is very close. Using the vertical lines associated with the correct scale makes this easier.

# MEASURING

To measure a line using the scale you need to look at the feet and inches. Place the zero on one end of a given line using the side of the scale the line was created with. If the opposite end of the line does not align with a true foot measurement, which rarely would happen, then slide it down to the closest foot and look back at the inch section of the scale. Count the marks, calculate the quantity each mark represents—remember that in the $3/4$" scale each mark was a half an inch while each mark is two inches at $1/8$" scale—and you have the full length of the line in feet and inches.

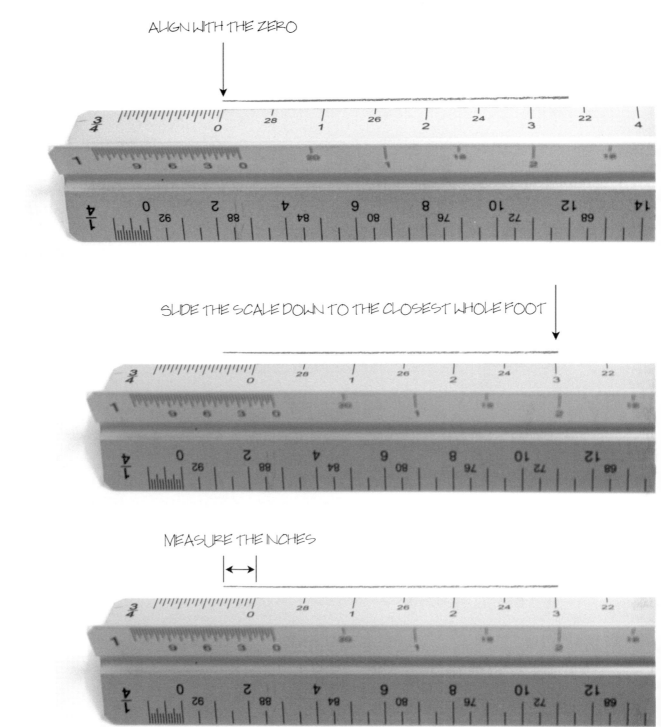

ALIGN WITH THE ZERO

SLIDE THE SCALE DOWN TO THE CLOSEST WHOLE FOOT

MEASURE THE INCHES

This line is 3'-4" at ¾" = 1'-0" scale.

# EXERCISE

|||||||||||||||||||||||||||||||||||||||||||||||||||||||||||||||||||||||||||||||||||||||||||||||||||||||||||||||||||||||||

## TOOLS NEEDED:

Architectural Scale                    Lead                              Lead Holder

Measure the following lines using the scale indicated.

1.  $\frac{1}{4}$" = 1'-0"        _____        ____' - ____"

2.  $\frac{3}{8}$" = 1'-0"        _____        ____' - ____"

3.  3" = 1'-0"        _____        ____' - ____"

4.  $\frac{1}{2}$" = 1'-0"        _____        ____' - ____"

5.  $\frac{1}{8}$" = 1'-0"        _____        ____' - ____"

# TIPS:

- Remember to align the end of the line with the nearest foot then read the inches.

- Try measuring each line with the opposing scale to see what length you get—if it is half and/or twice the length, you have measured it correctly.

# EVALUATION:

- Review the final measurements with your classmates. In the case of understanding the scale, sometimes hearing another person explain it to you can make all the difference.

# TEMPLATES

A template is used to draw predetermined shapes or symbols that are typical to architectural drawings. Shape templates can include circles, ellipses, and rectangles and have a progression of the shape in the actual drafted size: so if you wanted to draw a half inch circle you would select the circle cut out labeled $\frac{1}{2}$".

Symbol templates can include plumbing, electrical, and furniture symbols. These templates are

typically scaled to either reflect a quarter of an inch equals a foot ($\frac{1}{4}$" = 1'-0") or an eighth of an inch equals a foot ($\frac{1}{8}$" = 1'-0"); both standard sizes for architectural drawings. The template will have the scale printed on it so you know with which scaled drawing it is intended to be used.

Like triangles, templates are designed to allow you to also draft using ink because they have small risers built in to avoid smudging.

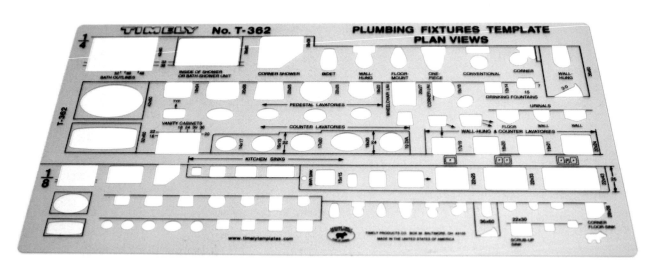

This template has basic plumbing symbols for $\frac{1}{4}$" = 1'-0" and $\frac{1}{8}$" = 1'-0" scale.

# COMPASS

|||||||||||||||||||||||||||||||||||||||||||||||||||||||||||||||||||||||||||||||||||||||||||||||||||||||||||||||||||||||||||||||||||||||||

The compass used in drafting is typically more durable than the ones you may have used previously and can expand to make greater arcs and circles with the addition of an arm. The compass comes loaded with graphite but can be removed to add an attachment that will allow you to use a pen or other instruments.

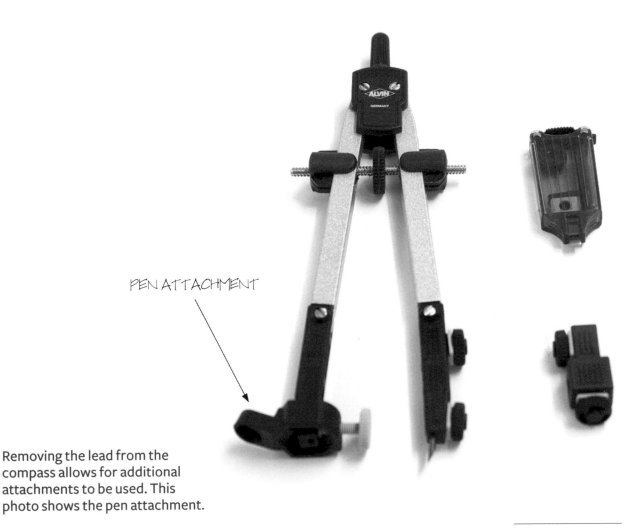

PEN ATTACHMENT

Removing the lead from the compass allows for additional attachments to be used. This photo shows the pen attachment.

# FRENCH CURVES
# AND FLEXIBLE CURVES

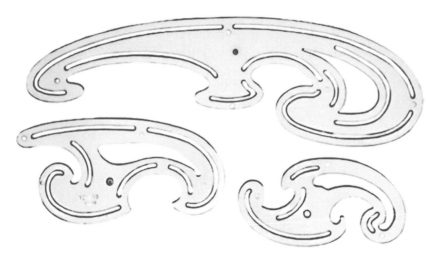

A set of French curves.

French curves are used to accurately make other types of curved lines. You rotate the French curve around until it fits the curve you wish to draw. Flexible curves can also be used to do the same thing but use caution as they can move easily while drawing.

# PAPER

The type of paper used for drafting varies but most people prefer vellum when drafting with lead. Vellum allows the lead to travel across the paper smoothly and evenly while still being forgiving enough to erase it if you make a mistake. Mylar is best used when drafting in ink. Both come in sheets, rolls, and pads and should be selected based on the drawings you are going to produce and the method in which you are going to produce them.

Other papers you might use when designing—not drafting—include tracing paper, grid paper, and bond paper. Tracing paper, typically used in rolls, comes in a variety of lengths and is available in yellow (buff) or white. Pads are also available but both are fairly transparent and will allow you to layer sheets on top of one another to build on an idea or drawing.

Various papers to use when drafting.

Grid paper can be used to lay out an idea quickly and to scale. Architecturally it is best to use the 4 × 4 grid which equates to $\frac{1}{4}$" = 1'-0" scale.

Bond paper is simply copy or printer paper that is good for getting an idea out quickly but it should never be used to draft.

# DRAFTING DOTS AND TAPE

You will need to secure your paper selection to your drafting board with tape, but not just any old tape, drafting tape. It is specifically made to attach to your board but also to lift off without damaging the corners of the paper.

Drafting tape is sold in rolls but it is also sold in much easier to use dots. Drafting dots dispense from a box by gently pulling on a tab while pressing lightly on the top where the dots are coming out.

Always use drafting dots, or tape, on all four corners so your paper stays secure and, although tempting, do not reuse them because they lose the tackiness after you remove them.

Gently pull the tab to expose a drafting dot, remove with your thumb, and attach to the corner of your paper.

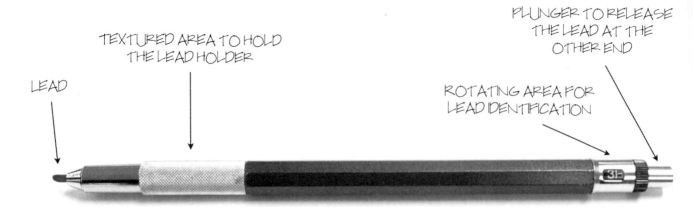

TEXTURED AREA TO HOLD
THE LEAD HOLDER

PLUNGER TO RELEASE
THE LEAD AT THE
OTHER END

LEAD

ROTATING AREA FOR
LEAD IDENTIFICATION

The anatomy of a lead holder.

# LEAD HOLDER, LEAD, AND LEAD POINTERS

To physically draw with lead using any of the tools previously discussed you will need a lead holder, lead, and a lead pointer.

The lead holder does exactly as the name implies, it holds the different leads used to draft. A standard lead holder has a plunger at the top, where a traditional eraser would be, that allows you to adjust how much lead is exposed. Below that is a small, rotating collar that has a window in it to reveal a letter and/or number. This refers to the lead type

that is loaded into the lead holder. This window should be rotated to the proper lead type anytime it is changed. At the bottom of the lead holder is a textured, typically metal, section. This is where you would hold the lead holder.

This may seem like a silly aspect of the lead holder but when you draft, accuracy is very important and having a slip-resistant surface aids in that accuracy. How to use the actual lead holder is discussed later in this chapter.

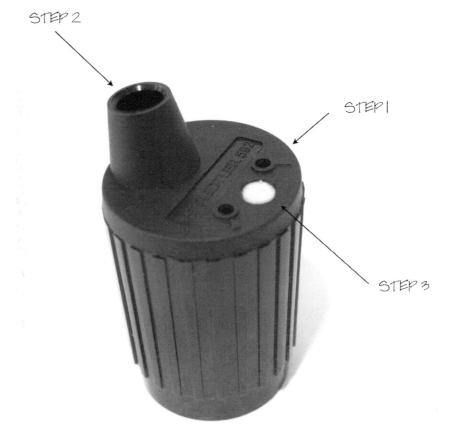

STEP 2

STEP 1

STEP 3

The anatomy of a lead pointer.

A lead pointer is used to sharpen the lead while in the lead holder. Simply place the lead holder above one of the holes on either side of the white dot and release the plunger (Step 1). This allows for the perfect amount of lead to be released to use the pointer. Next place the lead, while in the holder, into the larger hole until you can feel the lead stop, then rotate it around the lead pointer while you hold it still on your drafting surface (Step 2). Listen for the sound of the lead being sharpened, when you no longer hear the sound it is ready to use. Once you take the lead and holder out, poke the end of the lead into the white circle, this will remove any graphite dust off your newly sharpened lead (Step 3).

Leads are simply graphite that is inserted into the lead holders. The leads vary in density and go from hard (H) to soft (B). The hardest is 9H and 6B is the softest. It is very important you let the lead do the work; there is no need to "muscle" a line. How to choose which lead to select is discussed in Chapter 2 as it relates to line weights.

Typical drafting pen.

# INK, PENS, AND MARKERS

Technical pens are also used for drafting, typically on mylar, and have a built-in benefit of coming in a variety of tip sizes know as nibs. The variety in size assists with creating line weights, which will be discussed in Chapter 2.

Pens are sold individually or in sets and come in refillable or disposable models. The disposable models are usually referred to as markers. Although they can get the job done, the refillable pens tend to work better and obviously last a lot longer. Using a special cleaning fluid extends the life of the refillable pens and replacement pen points are also available should a nib become damaged.

# ERASERS AND ERASING SHIELDS

||||||||||||||||||||||||||||||||||||||||||||||||||||||||||||||||||||||||||||||||||||||||||||||||||||||||||||||||||||||||||||||||||||||||

Several handheld erasers have probably been included in your kit and you select which one to use based on what you are trying to erase and on what type of paper. Each is labeled for clarity.

You can use an erasing shield in any situation to get into a drawing more accurately, while shielding the other areas through a variety of shapes and sizes. Always use the next tool, a drafting brush, to remove the eraser pieces to ensure your drawings stay smudge free.

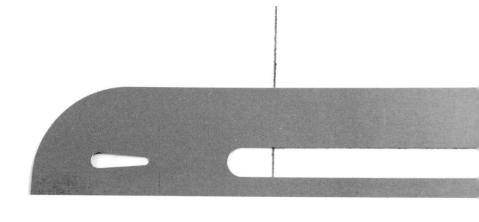

Lay the erasing shield over the line you want to erase and use the proper eraser to remove the line.

# DRAFTING BRUSHES
# AND DRY CLEANING PADS

Drafting can be a messy undertaking so these two tools are used to keep things neat and tidy. Keep the drafting brush and dry cleaning pad, usually referred to as pounce, on your drafting surface at all times. If you are right-handed keep the brush on the left and the pounce on the right and the opposite if you are left-handed.

Drafting brushes and dry cleaning pad and powder.

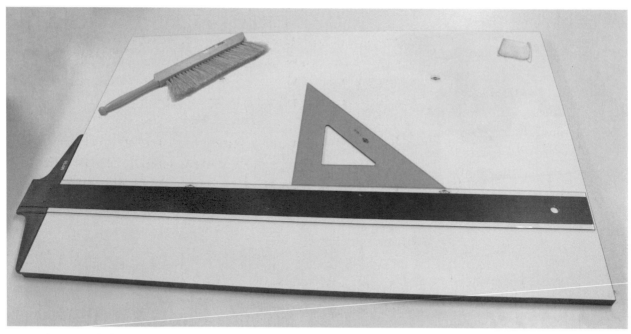

Keep these tools handy to keep drafting area and drawings clean.

When you start a drawing you should take the pounce bag and run it through your fingers above the paper you are going to draw on. It will release grit onto the paper that allows your tools to glide more freely as you draft. After a few minutes you should take the drafting brush and sweep away the grit, since it has collected the graphite, and apply the pounce again. As with many things in drafting, this will become second nature as you draft more often.

# SUMMARY

Learning the names and uses of your drafting tools sets you up to create beautiful drawings to communicate your design ideas. The following chapters will address the lines, types of drawings, and lettering you will use to do just that.

Finally, you should put your initials on each tool or somehow mark them to identify them as yours. Some people put nail polish on each tool or use colored electrical tape, but no matter what the method, make sure it is for you exclusively.

# TWO
# LINES

## OBJECTIVES

You will be able to identify and successfully use:

- The basic tools for technical drafting
- The correct line types and weights
- The graphic components of a drawing

DRAWINGS ARE USED TO COMMUNICATE YOUR DESIGN INTENT AND ALL DRAWINGS ARE COMPOSED OF LINES. THE USE OF SPECIFIC LINE TYPES AND WEIGHTS AIDS IN THE CLARITY OF THE DRAWINGS. REMEMBER YOU ARE NO LONGER JUST USING WORDS TO SELL AN IDEA, YOU ARE ALSO USING DRAWINGS AS WELL. THE SUCCESS OF THOSE DRAWINGS SPEAKS MUCH LOUDER THAN WORDS.

# HOW TO DRAW A LINE

This seems very simple but did you realize that your entire body goes into drawing a line? Sit at your drafting board or desk with both feet firmly on the floor. Make sure your seat is comfortable and at the right height for you to see the entire paper on your board. Take your lead holder into your hand by grabbing the textured portion near the bottom with just your pointer finger and thumb evenly spaced apart. Then gently rest it on your middle finger, placing your pinkie on the paper.

Note where the thumb and pointer finger begin at the start of the twist and where they are at the end.

Once you have held the lead holder like this, and I know it feels awkward, practice twisting the lead holder—just in the air for now. Keep your elbow on the table and rotate the entire lead holder between your thumb and pointer finger. It should roll down the pad of your thumb while rolling up your pointer finger. Do this a few times taking note of the distance you have traveled at the end of the twist. Try slowing it down and go farther.

Also, concentrate on your shoulder, not your wrist, remembering that your entire body goes into drawing a line. Once all of this feels comfortable you are ready to try to draw an actual line.

Start with the easiest lines to draw: straight lines using your drafting tools, the T-square for horizontal lines, and triangles for vertical and angled lines.

# EXERCISE

||||||||||||||||||||||||||||||||||||||||||||||||||||||||||||||||||||||||||||||||||||||||||||||||||||||||||||||||||||||

## TOOLS NEEDED:

T-square

Triangles

Dry Cleaning Pad

Drafting Board

Leads

Lead Holder

Vellum

Drafting Dots

Drafting Brush

Lead Pointer

Set up your drafting area and prepare to draw a series of lines.

1. Tape down a piece of 8 ½ × 11 vellum using your T-square to line the paper up on the drafting board.

   - Note that you do not want to align the bottom of your paper with the bottom of your drafting board. Give yourself a few inches above it to ensure the successful use of your drafting tools.

2. Use drafting dots to secure the paper in place on all four corners.

3. Place the T-square near the top of the page.

4. Rest the lead against the top of the T-square and begin your line going from left to right.

5. Slide your T-square down the drafting board and continue to draw lines, striving to twist further, creating a longer line with each attempt.

6. Once you are comfortable with drawing horizontal lines try creating vertical and angled lines using your triangles.

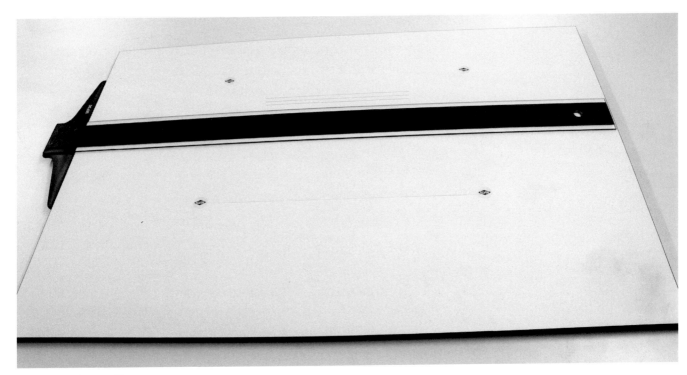

Set up to draw a line.

## TIPS:

- Make sure the T-square is "locked and loaded" against the drafting board.
- Twist the lead as you draw.

## EVALUATION:

- Did your lines get longer with each attempt?
- Can you tell, just by looking, when you did or did not twist the lead?
- Are all the horizontal lines parallel or did they start to angle? If so, then either your T-square wasn't "locked and loaded" or you did not keep the lead consistently on top of the T-square.

# LINE TYPES

Once you feel comfortable using the tools to draw lines we can add another aspect to your drafting skills by looking at the different line types. There are six standard lines used in drafting:

- Solid lines
- Dashed or hidden lines
- Centerlines
- Leader lines
- Dimension lines
- Guidelines

Each can have a darkness or lightness to them, referred to as the line weight, which further enhances the graphic success of a drawing. Line weights will be discussed later in the chapter.

Solid lines.

Dashed lines.

## SOLID LINES

Solid lines are used to define an edge of an object. In architectural drawings that can include a wall, floor tile, chair, etc., and vary in its use within a drawing.

## DASHED OR HIDDEN LINES

Architecturally dashed or hidden lines are used for items not in full view of where the drawing was taken. You draw short lines separated by spaces to create a dashed line. The length of the lines and spaces should be consistent for greater graphic understanding. However, the lines and spaces do *not* need to be the same measure.

Centerlines.

Leader lines.

## CENTERLINES

Centerlines are used to indicate the axis of a column or wall opening. They consist of a long line, a small space, a dash, then another small space, followed by a long line again.

## LEADER LINES

Leader lines are solid lines with an arrowhead on one end. The arrow points to the item being identified and the open end is where a note or comment should be written.

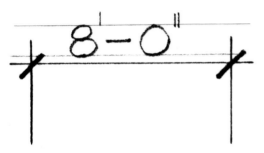

Dimension lines.

Guidelines.

## DIMENSION LINES

Dimension lines are solid lines with back slashes, referred to as terminators, on each end at a perpendicular line. The perpendicular lines align with a portion of a drawing while the back slashes cross through the intersection. A number in feet and inches is centered between the terminators, slightly above the solid line. Dimension lines are rarely used in presentation drawings but are seen in construction drawings.

## GUIDELINES

Guidelines are very light solid lines used to lay out a drawing or align lettering.

# LINE WEIGHTS

To enhance any type of line in your drawings you can add line weights. There are four basic line weight categories; however, there are variances within each:

- Heavy
- Intermediate
- Light
- Very Light

The more line weights you use the stronger a drawing will communicate your design ideas. It is important to understand when using line weights that darker lines advance while lighter lines recede in the drawing. This means a darker line will be seen by the viewer first therefore will communicate the strongest in a drawing.

An example of basic line weights.

# HEAVY

Heavy lines are the darkest of all the line weights, referred to as profile lines, and are typically drawn as a solid line. They are used to indicate major spatial edges including:

- Walls
- The definition of the thickness of the cut items on plan and section drawings
- The ground in an elevation or section drawing

You typically use H, F, HB, and B leads to create heavy lines. You will have to experiment by drawing lines to determine what is the best lead for you. If you are not getting the desired effect you may want to use a softer lead. The downside is it leaves more graphite on the paper, which can be dragged across and create smudges. Make sure to always use the dry cleaning pad and drafting brush when creating heavy lines.

# INTERMEDIATE

Intermediate lines are the middle line weight and are typically solid or dashed. They are used to indicated secondary information:

- Door and windows
- Cabinetry
- Plumbing fixtures and appliances
- Lettering
- Furniture

Use H, F, and HB leads to create intermediate lines.

# LIGHT

Light lines are the last visible lines on a drawing that will show up on a copy of the drawing. They are used to suggest materials, color, texture, or minor lines on a drawing:

- Door swings
- Tile pattern on a floor
- Chair rail on a wall

Use 2H, H, and F to create light lines. When using harder leads you can tear your paper, so remember to let the lead do the work for you.

# VERY LIGHT

Very light lines should be barely visible on the drawing and rarely show up on a copy. They are used for:

- Layout drawings
- Layouts for sheets
- Guidelines for lettering

Use the hardest lead you can handle to create very light lines—4H, 2H, or H.

# LINE INTEGRITY

||||||||||||||||||||||||||||||||||||||||||||||||||||||||||||||||||||||||||||||||||||||||||||||||||||

Making sure you create a line with purpose is line integrity. A line should have a defined beginning, unified and consistent middle, and established end. When lines meet or cross they should do so with a crisp exactness. A line is not worth drawing if it does not have integrity.

# EXERCISE

||||||||||||||||||||||||||||||||||||||||||||||||||||||||||||||||||||||||||||||||||||||||||||||||||||||||||||||||||||||||||||

## TOOLS NEEDED:

T-square
Lead Holder
Vellum
Triangles

Leads
Drafting Brush
Lead Pointer

Drafting Dots
Dry Cleaning Pad
Drafting Board

Set up your drafting area like in the first exercise and prepare to practice your line types and weights.

1.  Draw three lines using your drafting tools with your hardest lead, demonstrating each line type:

    - Solid lines

    - Dashed or hidden lines

    - Centerlines

    - Leader lines

    - Dimension lines

    - Guidelines

2.  Write the name of the lead you are using at the end of each line.
3.  Draw three more of each line using each of the leads you have in your kit.

## TIPS:

- Twist your lead as you draw.
- Pay attention to the integrity of each line.
- Sharpen your lead with your lead pointer often.
- Use the dry cleaning pad and drafting brush often.

## EVALUATION:

- Which lead created which line weight for you? Did it vary based on the line type?
- Which line type did you draw the best? Worst? Most consistently?
- Looking at the entire sheet, can you tell when you sharpened the lead?
- Again, looking at the entire sheet, can you tell when you did or did not twist the lead?
- Did your lines improve with each lead change?

# POCHÉ AND HATCHING

The use of poché and hatching, in addition to line types and weights, will enhance your drawings and make them graphically strong.

## POCHÉ

Poché is basically to completely color in between two lines; the lines on either side of a wall in a plan, for example. Always do the poché as the final step in creating a drawing because it can be messy. There are two schools of thought on how to actually poché.

Poché with colored pencil.

Poché with graphite.

1. Color in between the lines on the back of the drawing. I recommend using a red colored pencil, specifically a Prisma brand colored pencil. Pochéing on the back of the drawing allows for line integrity on the front of the drawing. The poché can be erased, if necessary, without jeopardizing the original integrity of the line.

2. Use powdered graphite, which can be purchased at a store, to color in between two lines. Place sticky notes along the edge of the lines on the drawing and, using a cotton swab, slowly put the graphite on in a circular motion. Once you are done pochéing with this method, the drawing should be sprayed with a product called spray fix to keep the graphite from smudging.

# HATCHING

Hatching is to fill in an area, or between two lines, with a pattern of lines. They can be drawn with tools or freehand. Each type of hatching can indicate a different height of a wall, for example. A legend may be used to denote what each hatching pattern indicates.

You can also create a hatch to simulate a tiled floor, for example, by drawing a grid, to scale, on a floor plan.

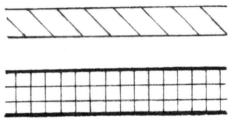

Hatching.

# SUMMARY

IIIIIIIIIIIIIIIIIIIIIIIIIIIIIIIIIIIIIIIIIIIIIIIIIIIIIIIIIIIIIIIIIIIIIIIIIIIIIIIIIIIIIIIIIIIIIIIIIIIIIIIIIIIIIIIIIIIIIIIIIIIIIIIII

The graphic success of your drawings is dependent on your ability to draw lines, give them the proper weight, and to add graphic enhancements to communicate your design intention. These basic drafting skills will serve you well throughout your career. The following is an example of a drawing before and after these principles were applied. Which do you think is more successful?

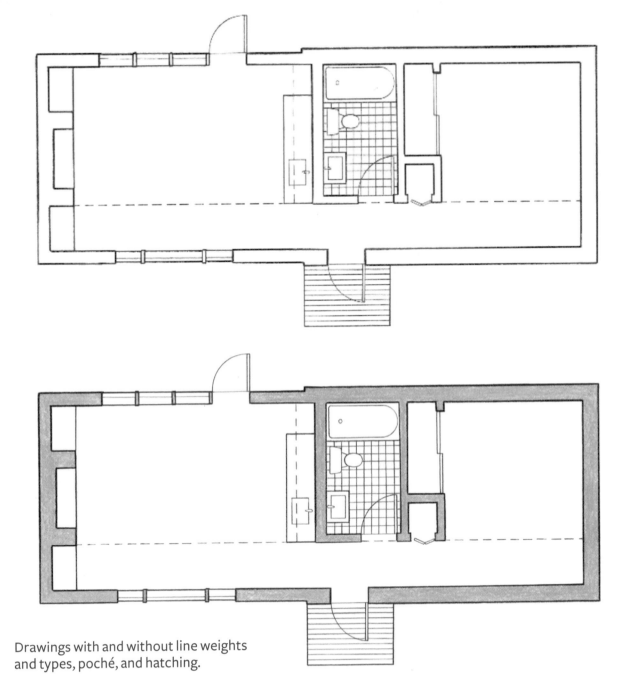

Drawings with and without line weights
and types, poché, and hatching.

# THREE
## ORTHOGRAPHIC DRAWINGS

### OBJECTIVES

You will be able to identify and successfully draw:

- Plans
- Elevations
- Sections

ORTHOGRAPHIC DRAWINGS ARE TWO-DIMENSIONAL, GRAPHIC REPRESENTATIONS OF THREE TYPES OF DRAWINGS KNOWN AS PLANS, ELEVATIONS, AND SECTIONS. THEY ARE ALL DRAWN TO A MEASURABLE SCALE USING DRAFTING TOOLS.

TO FULLY UNDERSTAND ALL OF THE ORTHOGRAPHIC DRAWINGS, AS WELL AS THE THREE-DIMENSIONAL DRAWINGS IN BOOK THREE OF THE STUDIO COMPANION SERIES, A SMALL ONE-BEDROOM CABIN WILL BE USED THROUGHOUT. IN ADDITION, HAVING THIS CABIN DOCUMENTED IN EVERY DRAWING TYPE WILL ASSIST YOU WHEN SELECTING WHICH IS BEST SUITED TO COMMUNICATE YOUR DESIGN INTENT.

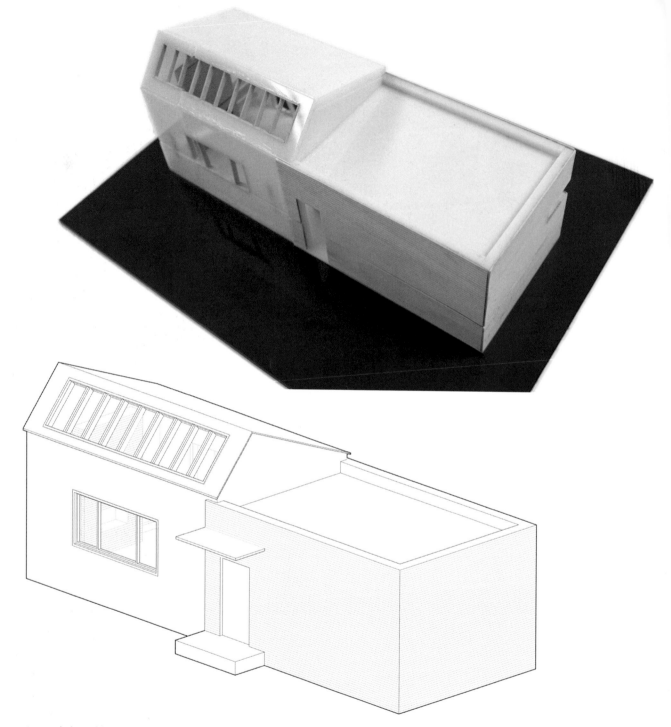

A model and isometric of the one-bedroom cabin.

# PLANS

||||||||||||||||||||||||||||||||||||||||||||||||||||||||||||||||||||||||||||||||||||||||||||||||||||||||||||||||

There are several types of plans that are used in our industry. Floor plans are the most common and the first type of plan students typically learn to draw. Floor plans are horizontal drawings of the space or building.

A floor plan indicates:

- Size and shape of the building and the overall *length* and *width* of the space
- Configuration of walls
- Location of columns, doors, windows, openings, and means of vertical circulation including stairs, ramps, and elevators
- All the items a contractor would build or install including plumbing fixtures and built-in cabinetry

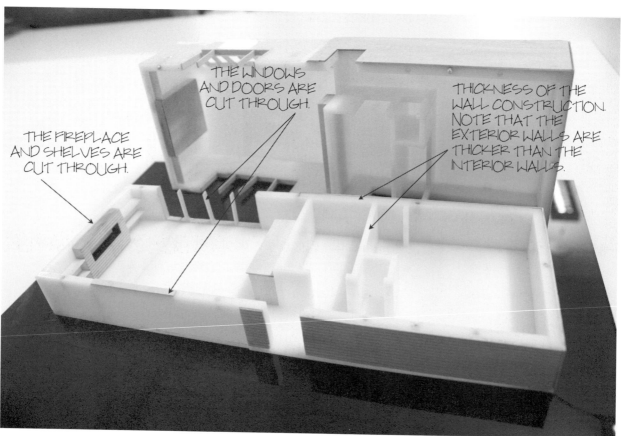

THE WINDOWS
AND DOORS ARE
CUT THROUGH.

THICKNESS OF THE
WALL CONSTRUCTION.
NOTE THAT THE
EXTERIOR WALLS ARE
THICKER THAN THE
INTERIOR WALLS.

THE FIREPLACE
AND SHELVES ARE
CUT THROUGH.

Anatomy of a floor plan.

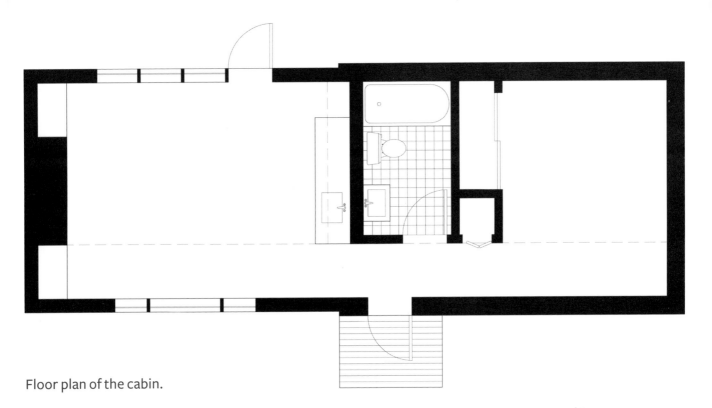

Floor plan of the cabin.

To create a floor plan:

- Visually cut the building approximately 3'-6" above the floor

- Represent everything that is cut through using a solid line

- Represent items higher than the cut with dashed lines—like
  a loft that overlooks the space from above or upper kitchen
  cabinets; dashed lines below a cut are smaller while those above
  the cut are longer

- Draw the plan to scale; typically at $\frac{1}{4}$" = 1'-0"

- Draw the plan so North is up on the sheet whenever possible

To draft a floor plan:

1.  Using guidelines and your drafting tools, establish the upper left corner of the plan by drawing the North and West walls.

2.  Working left to right, draft all the vertical lines that represent all the North/South walls in the building. Once all the vertical lines have been drawn use the same method to create all the East/West walls working top to bottom.

3.  Continue using guidelines to locate items, to scale, within the plan.

    ▪ Doors, windows, stairs, etc.

    ▪ Plumbing fixtures and cabinetry

    ▪ Flooring

4.  On a clean piece of paper trace over the guidelines with your leads using the correct line types and weights.

    ▪ Until you are comfortable with line weights, make a copy of your plan and use highlighters or colored pencils to indicate which line weights belong in the drawing, then do the final drawing.

5.  Finally, poché the thickness of walls and structure.

    ▪ Remember we poché the walls because we want to indicate the thickness of the construction of the building. The wall assembly at $1/4'' = 1'-0''$ scale is typically too small to accurately illustrate. Architectural details will be drawn for the construction of the project at a much larger scale.

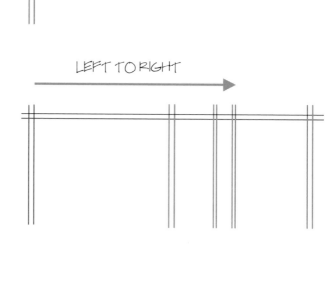

LEFT TO RIGHT

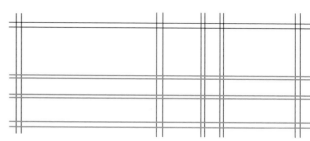

TOP TO BOTTOM

**Draft each line using your tools.**

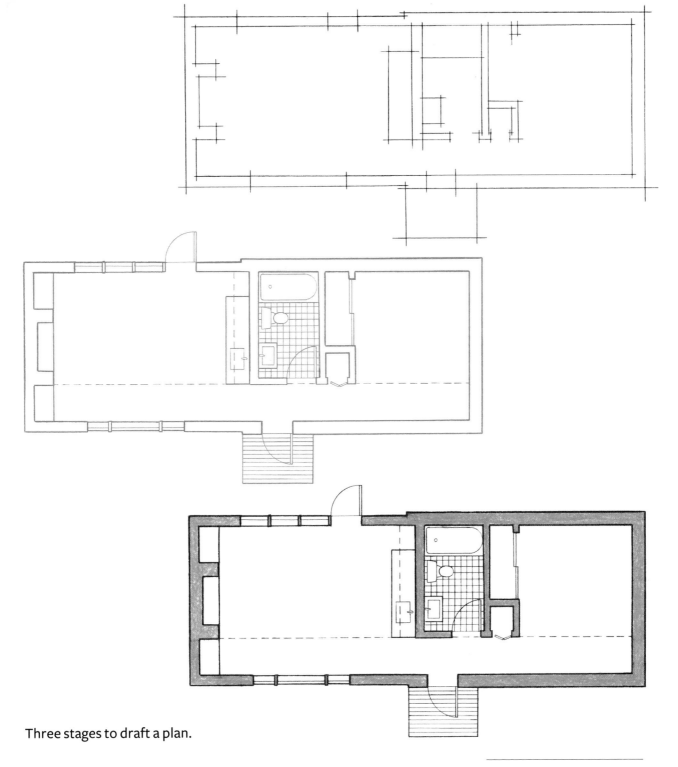

Three stages to draft a plan.

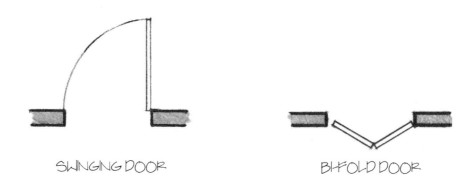

SWINGING DOOR                    BI-FOLD DOOR

SLIDING DOOR/WINDOW

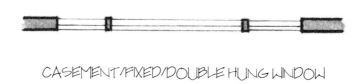

CASEMENT/FIXED/DOUBLE HUNG WINDOW

An enlarged drawing of each
door and window type used
in the cabin. Each is drawn
to represent the method or
motion they take to operate.

Some other types of plans include:

- Site plan—indicates the land and surroundings of the building
- Finish plan—indicates the materials and finishes
- Furniture plan—indicates the location of all the freestanding furniture
- Lighting plan or reflected ceiling plan (RCP)—indicates all the light fixtures, switching, ceiling heights, and elements
- Mechanical, plumbing, and electrical plans—created by engineering consultants with information selected by the design team

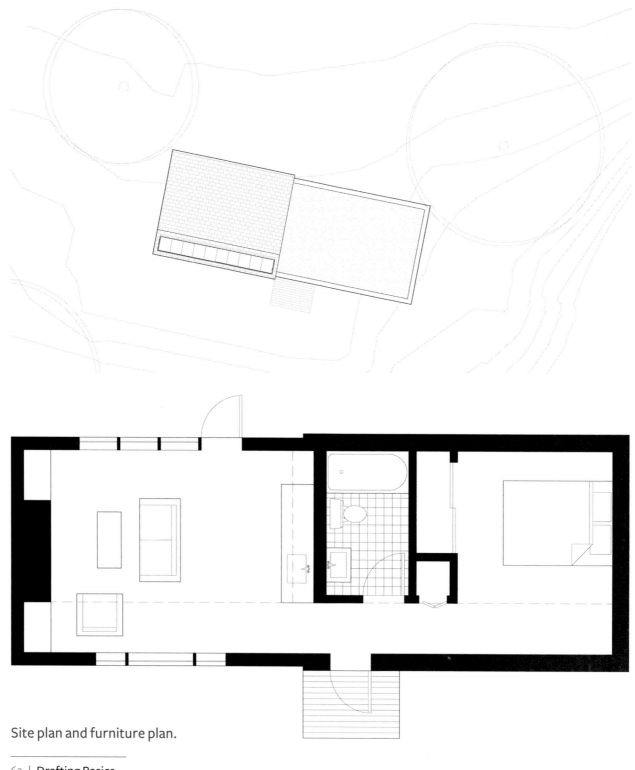

Site plan and furniture plan.

# ELEVATIONS

||||||||||||||||||||||||||||||||||||||||||||||||||||||||||||||||||||||||||||||||||||||||||||||

Elevations are vertical drawings typically of interior walls or exterior sides of a building. An elevation looks exactly the same no matter where you stand in the space. It looks the same if you are two feet away or twenty feet away.

An elevation indicates:

- The *height* of the overall space
- The heights of doors, windows, and openings
- Materials and finish locations
- All the items a contractor would install or build, including plumbing fixtures and built-in cabinetry

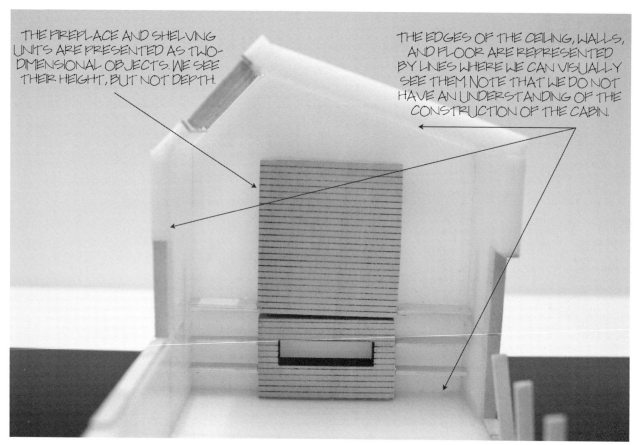

THE FIREPLACE AND SHELVING UNITS ARE PRESENTED AS TWO-DIMENSIONAL OBJECTS. WE SEE THEIR HEIGHT, BUT NOT DEPTH.

THE EDGES OF THE CEILING, WALLS, AND FLOOR ARE REPRESENTED BY LINES WHERE WE CAN VISUALLY SEE THEM. NOTE THAT WE DO NOT HAVE AN UNDERSTANDING OF THE CONSTRUCTION OF THE CABIN.

Anatomy of an elevation.

To create an elevation:

- Typically drawn at $\frac{1}{4}$" = 1'-0" or the same dimension as the plan
- Draw everything you can see on the wall with your eyes
- Elevations are as wide and tall as the room you are standing in and nothing more

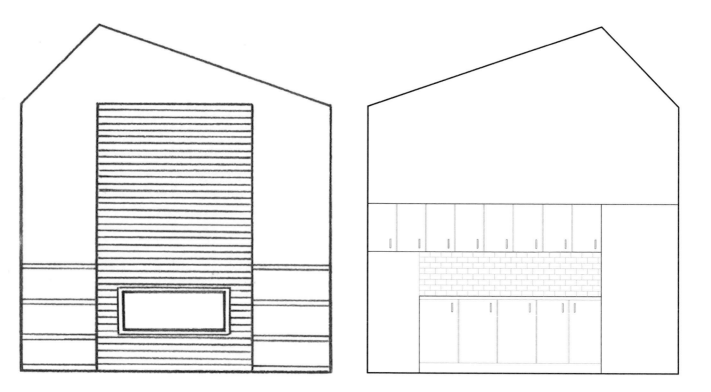

The East elevation of the fireplace wall and the West elevation of the kitchen wall on the interior of the cabin.

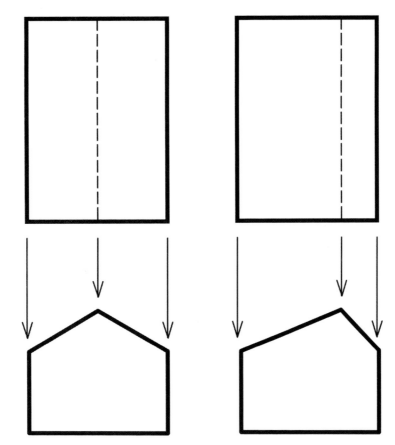

**Using a floor plan, with a dashed line that indicates the top of the slope of the ceiling, allows for that location to be pulled, or extended, down for the elevation.**

To draft an elevation:

1. Tape a copy of the floor plan above where you want to draw the elevation. Remember to draw the elevation as if you are standing *in* the room.

2. Using your drafting tools, draw a horizontal guideline line all the way across the page about 1" above the bottom edge of the paper.

   - This will be the floor of the space.

3. For a typical interior space, draw another line 8'-0" above that line now using the ¼" = 1'-0" scale.

   - This will give you the height, or ceiling line, of the space, if the room has a twenty foot ceiling then the second line would be drawn 20'-0" above the first line using the ¼" = 1'-0" scale.

   - If the ceiling is angled, you will need to locate the height of the walls where they meet the sloped ceiling and find the high point of the ceiling to connect the two.

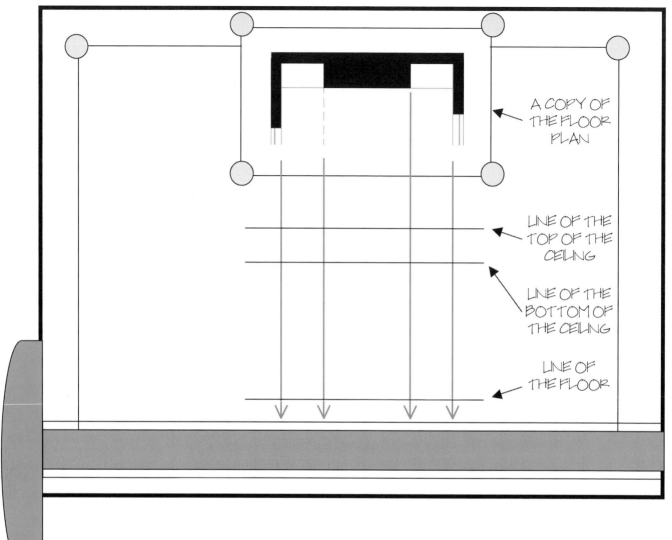

A COPY OF THE FLOOR PLAN

LINE OF THE TOP OF THE CEILING

LINE OF THE BOTTOM OF THE CEILING

LINE OF THE FLOOR

Draft each line using your tools.

4. Begin by using your triangle to pull down, or extend, each vertical line from the plan. This would include the edges of any window, door, or opening that will be needed to create the elevation using very light guidelines.

5. Establish the heights of any doors, windows, materials, and finishes by measuring up from the ground line using your scale, then draw a horizontal guideline.

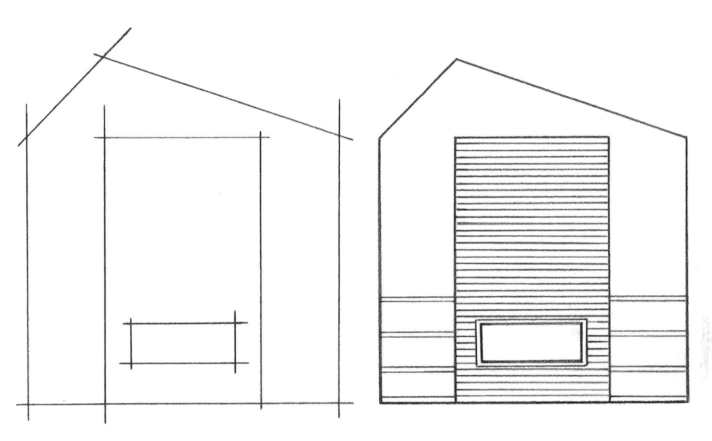

Two stages to draft an interior elevation.

6. Once you have all the guidelines use the proper leads and line weights to go over them to create the final elevation.

7. Remove the floor plan and rotate it so you can create the next elevation using the same steps.

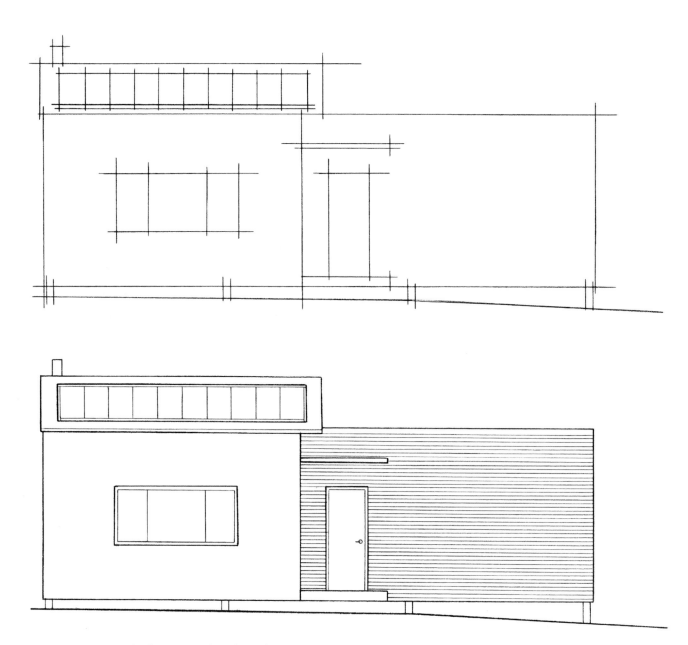

Two stages to draft an exterior elevation.

# SECTIONS

||||||||||||||||||||||||||||||||||||||||||||||||||||||||||||||||||||||||||||||||||||

Like elevations, sections are vertical drawings; however, they are cut through the entire structure and indicate the construction thickness of the walls, floors, and ceilings.

A section indicates:

- The building relationships from floor to floor and wall to wall
- The construction of the building

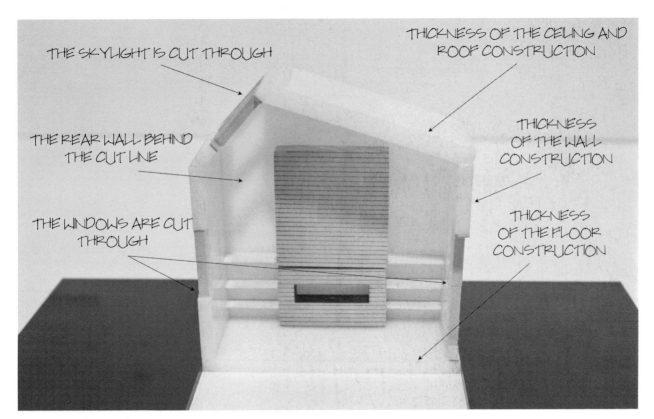

THE SKYLIGHT IS CUT THROUGH

THICKNESS OF THE CEILING AND ROOF CONSTRUCTION

THE REAR WALL BEHIND THE CUT LINE

THICKNESS OF THE WALL CONSTRUCTION

THE WINDOWS ARE CUT THROUGH

THICKNESS OF THE FLOOR CONSTRUCTION

Anatomy of a section.

A section is cut where you want to see the construction assembly as well as the elevation. In this photograph the section is cut about 6'-0" from the fireplace wall. Once the model is separated we can get the West elevation of the fireplace and the East elevation of the kitchen as well as the construction or thickness of the walls, floor, and ceiling.

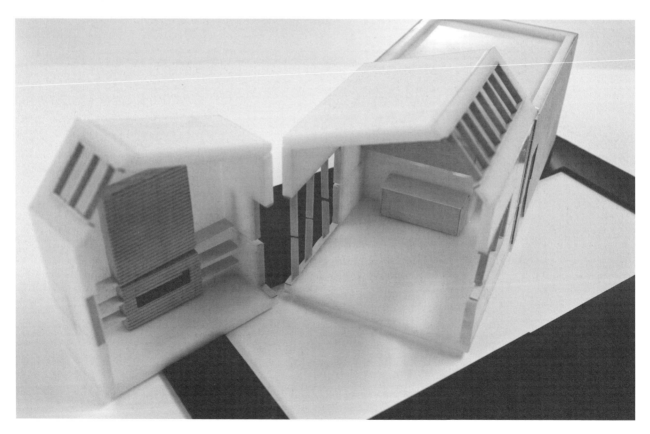

To create a section:

- Visually cut all the way through the building where you want to learn about the building relationships and determine which direction to look at the cut location so you gain the most information

- Represent everything that is cut through using a solid line

- Sections are as wide and tall as the entire building assembly and site, as if you have X-ray vision

- Draw everything that is cut through

    - Floors, walls and ceilings as well as items such as the glass and frame of a window

- Draw the elevation at the back of the space on each floor as required

- Draw the section to scale; typically at $1/4" = 1'-0"$

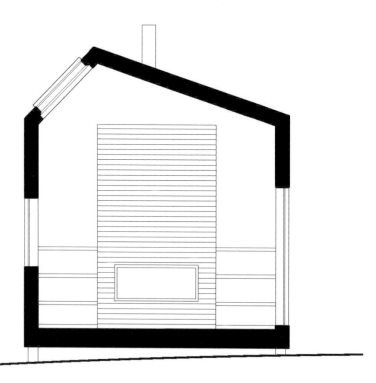

Each section indicates the elevations seen in the direction that the section is being cut. On the left is a section looking toward the West wall, indicating the heights of the windows on each sidewall as well as the elevation of the fireplace. Below is a section that reveals the relationship between the living/kitchen area to the bathroom, closet, and bedroom as well as the ceiling and floor construction heights and thickness. In addition you see the North elevations of each room.

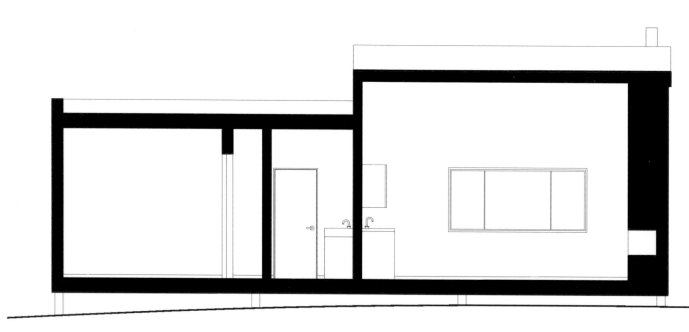

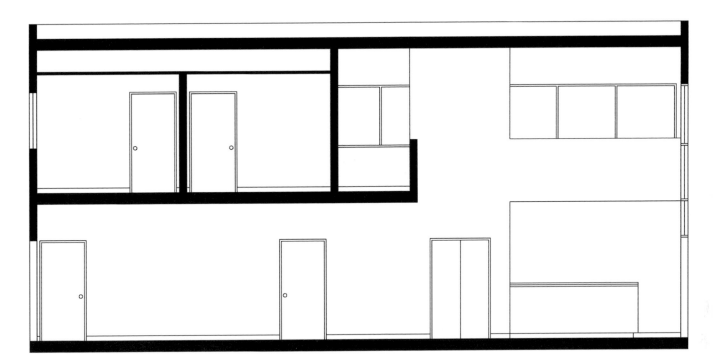

As with an elevation you can locate the heights of the different rooms and items on the elevation and you can also see the relationships between floors and rooms. In this drawing you can see the two floors of the building as well as the two-story lobby space on the right.

To draft a section:

1.  Just like with an elevation, tape a copy of the floor plan above where you want to draw the section and begin to "pull down" each line.

    - In addition to all the lines you pulled for an elevation, you will also pull the lines from the outside of the building and anything else you cut through, like the glass and frame of a window.

2.  Using your drafting tools draw a horizontal guideline line all the way across the page near the bottom edge of the paper.

3.  Beginning with the first-floor space, draw another line 8'-0" above that line now using the $1/4$" = 1'-0" scale.

    - As with the elevations, this will vary depending on your specific space. Treat each space separately and follow the methods for the elevations to draw the heights within the space.

4.  Establish the heights of each floor and ceiling as well as any doors, windows, materials, and finishes by measuring up from the ground line, using your scale, then draw a horizontal guideline.

5.  Once complete, draw the thickness of the floors, ceilings, and anything else within their assembly.

6.  Poché the thickness of walls, floors, and ceilings.

    - We poché because we want to indicate the thickness of the construction of the building.

    - We do not indicate the wall assembly because at $1/4$" = 1'-0" scale it is too small to accurately illustrate it. Architectural details will be drawn for the construction of the project at a much larger scale.

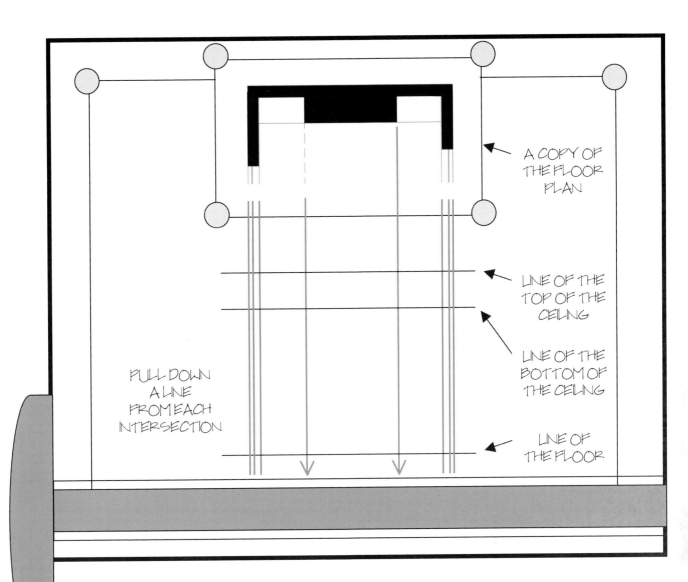

A COPY OF
THE FLOOR
PLAN

LINE OF THE
TOP OF THE
CEILING

LINE OF THE
BOTTOM OF
THE CEILING

LINE OF
THE FLOOR

PULL DOWN
A LINE
FROM EACH
INTERSECTION

Draft each line using your tools.

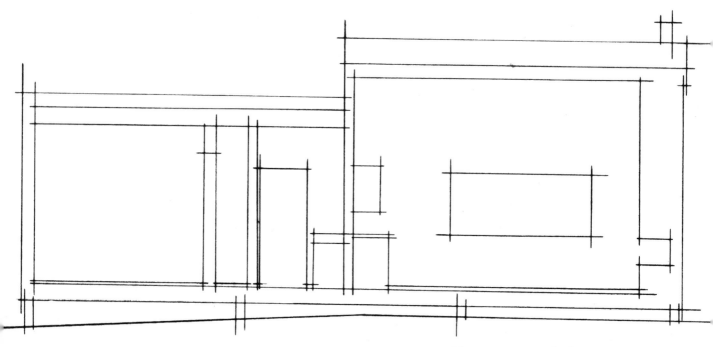

Three stages to draft the East section.

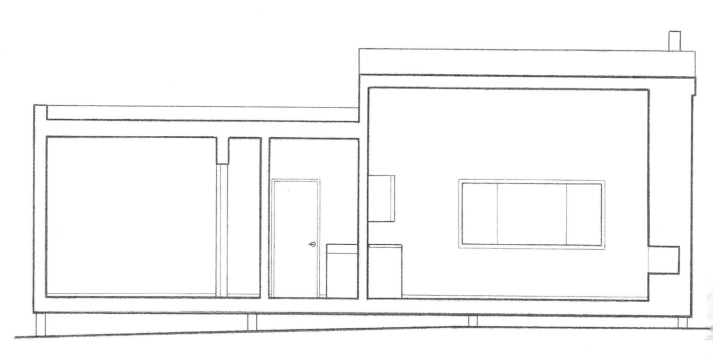

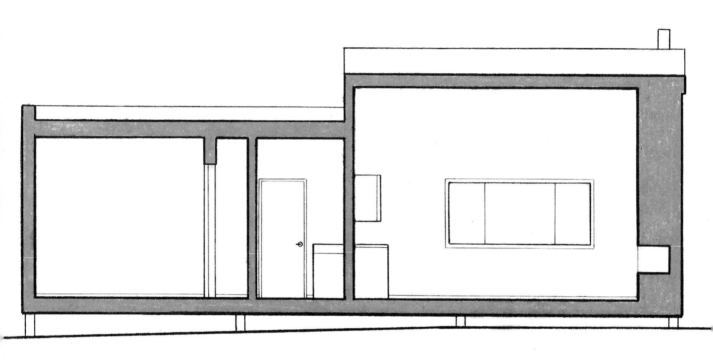

# SUMMARY

||||||||||||||||||||||||||||||||||||||||||||||||||||||||||||||||||||||||||||||||||||||||||||||||||||||

Orthographic drawings include plans, elevations, and sections and when drawn properly they communicate your design intent. Additional drawings may be needed to further explain your design— these drawings are typically three-dimensional and the third book in the Studio Companion Series addresses these efforts.

# FOUR
# ARCHITECTURAL LETTERING

## OBJECTIVES

You will be able to create:

- Quality architectural letters
- Successful word and sentence spacing

YOU HAVE BEEN WRITING YOUR ALPHABET FOR TOO MANY YEARS TO COUNT; HOWEVER, YOU WILL HAVE TO LEARN TO WRITE LIKE AN ARCHITECT AND DESIGNER. SOME PEOPLE FEEL LEARNING TO DO ARCHITECTURAL LETTERING IS A RITE OF PASSAGE. ONCE YOU LEARN TO LETTER THIS WAY YOU WILL NEVER GO BACK AND PEOPLE WILL SEE YOU AS AN ARCHITECT EVEN WHEN YOU ARE WRITING A CHECK OR MAKING A SHOPPING LIST.

Before computers, all architects had to write the same way to maintain consistency as multiple people worked on a project. As you will recall from Chapter 2, all designers use the same line types and weights to make drawings consistent, it is the same way with lettering.

Architectural letters are based on vertical, horizontal, and some angled and curved strokes. Each letter is created with some simple yet strict rules to follow while you are initially learning. Once you have mastered this strict method you will start to develop your own signature style.

# GUIDELINES AND LETTERING GUIDE

The golden rule of lettering is to use guidelines. There is a tool in your kit that was not discussed in Chapter 1, the Ames Lettering Guide. This little piece of plastic will save you a ton of time and energy when it comes to creating guidelines. It aids in drawing guidelines quickly, allowing for a consistent letter height while also keeping your words and sentences straight on the page.

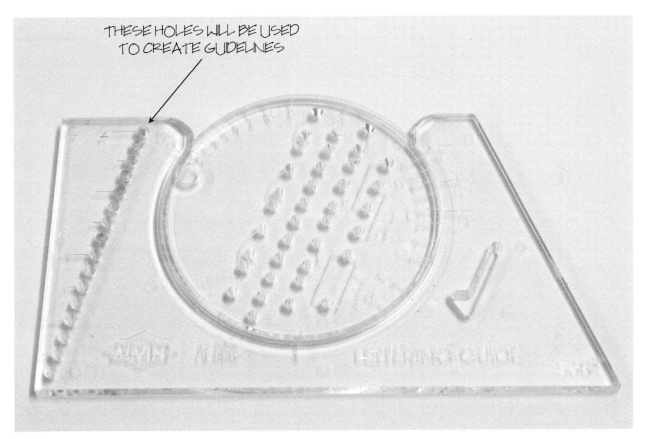

THESE HOLES WILL BE USED TO CREATE GUIDELINES

The lettering guide is used to create guidelines and strokes for architectural lettering.

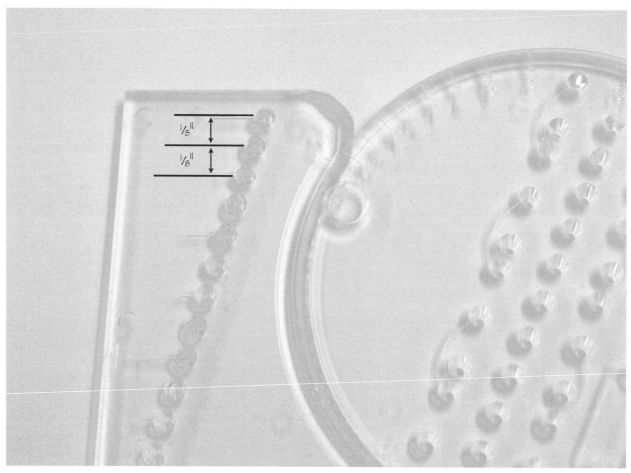

Each hole allows you to create a $1/8$" space between guidelines for lettering.

## USING THE LETTERING GUIDE

On the left side of the lettering guide you will see a series of 16 holes along an angled line. Each of these holes are spaced $1/8$" apart and will allow you to create, in essence, the paper you probably first learn to letter on in kindergarten. Do you remember that paper with blue solid lines followed by the red dashed line then another blue solid line? The holes on the lettering guide basically will allow you to create that type of paper anywhere you want on a sheet.

EDGE OF
T-SQUARE

A series of guidelines are created using the lettering guide.

Place the lettering guide on top of the T-square and slide it like you would a triangle. Gently put the point of your lead into the top hole and support the lettering guide with your pinkie finger as it travels across the T-square. This will take some practice but, just like riding a bike, once you finesse it you will be able to draw guidelines all day long.

Once you have created your first guideline, leave everything exactly where it is—do not move the T-square or lettering guide. Simply move the lead into the hole below the one you just used and travel back across the page. Continue to do this until you reach the bottom hole, move the T-square down the board, place your lead into the top hole and align it with the last guideline you drew and start again. Repeat this process until you have enough guidelines to begin lettering.

## HOW TO DRAW EACH LETTER

As stated before each letter is created through a series of strokes. The order in which a stroke is written is important to the success of the individual letter. The alphabet on the inside back cover is color coded to the step-by-step order in which the strokes should be created. Pink is the first stroke, blue second, orange third, and finally green.

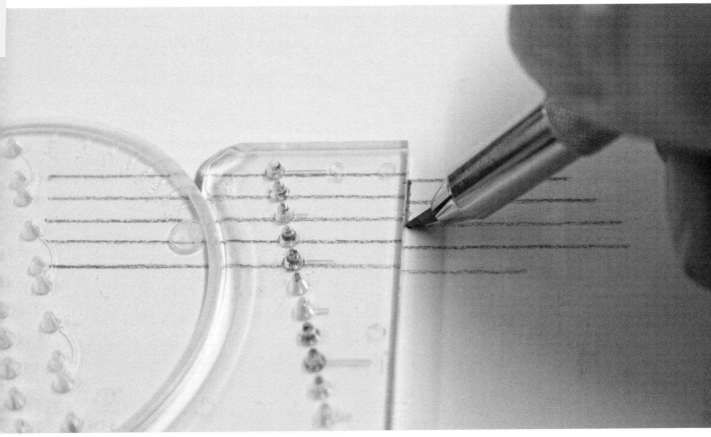

A horizontal stroke is created using the edge of the lettering guide.

Vertical strokes should be from the top down and drawn using a small triangle or the straight edge of the lettering guide when flipped over.

Horizontal, angled, and curved strokes should be drawn freehand. It may take a few tries to correctly get the letter but try to draw each letter exactly how it is shown. You may notice that the first two strokes typically establish the width of the letter.

When completing, for example, an "M" the angled strokes may not line up in between perfectly but this is why we practice our lettering. Inevitably you will develop your own style once you have mastered these rigid letters. Now that I have been lettering for over twenty years, I typically draw all the horizontal strokes within a letter at a small, upward angle.

The guidelines are used to keep the rows of letters straight across the page as well as providing a consistent space between lines of text.

Typically your letters should be $\frac{1}{4}$" high but can vary for presentation reasons. More lettering heights and styles will be discussed in book four of the Studio Companion Series but for now concentrate on $\frac{1}{4}$" high letters. The guidelines that you created with the lettering guide provided a line every $\frac{1}{8}$". Use the middle line to assist with any horizontal strokes needed for a letter.

# EXERCISE

||||||||||||||||||||||||||||||||||||||||||||||||||||||||||||||||||||||||||||||||||||||||||||||||||||||||||||||||||||||||||||||||||||

## TOOLS NEEDED:

T-square

Lead Holder

Lettering Guide

Drafting Board

Leads

Drafting Brush

Vellum

Drafting Dots

Dry Cleaning Pad

Lead Pointer

Set up your drafting area and prepare to draw guidelines and your architectural letters.

1. Tape down an 8½ × 11 piece of vellum using your T-square to line the paper up on the drafting board.
2. Use the drafting dots to hold the paper in place on all four corners.
3. Place the T-square near the top of the page.
4. Rest the lettering guide against the top edge of the T-square and begin to draw your guidelines, moving the T-square as needed, until you fill the page.
5. Flip your lettering guide over to use the straight edge to create the vertical strokes.
6. Create each letter, stroke by stroke—if you are not satisfied with the letter draw it again.
7. Create each number, stroke by stroke.

## TIPS:

- Make sure the T-square is "locked and loaded" against the drafting board at all times.
- Don't worry if your letters look a little strange—too big, tilting, uneven—that is why you are practicing.
- Once you are comfortable with each letter and number start writing words and then graduate up to sentences.
  - Letter your class notes or a favorite quote as practice.

## EVALUATION:

- Did you always use your straight edge for the vertical strokes?
- How successful were the curved letters and numbers?
- How confident were you with your strokes—can you see improvement on the sheet?
- How consistent is the spacing between letters and words?
- How well did you draw the symmetrical letters?

# SUMMARY

||||||||||||||||||||||||||||||||||||||||||||||||||||||||||||||||||||||||||||||||||||||

Architectural lettering is an integral part of a successful drawing and graphic consistency is essential in anything you write. Practicing will only make your lettering stronger and make it part of your everyday life.

# BASIC METRIC
# CONVERSION TABLE

## DISTANCES

| ENGLISH | METRIC |
|---|---|
| 1 inch | 2.54 centimeters |
| 1 foot | 0.3048 meter / 30.38 centimeters |
| 1 yard | 0.9144 meter |

| METRIC | ENGLISH |
|---|---|
| 1 centimeter | 0.3937 inch |
| 1 meter | 3.280 feet |

## WEIGHTS

| ENGLISH | METRIC |
|---|---|
| 1 ounce | 28.35 grams |
| 1 pound | 0.45 kilogram |

| METRIC | ENGLISH |
|---|---|
| 1 gram | 0.035 ounce |
| 1 kilogram | 2.2 pounds |

# GENERAL FORMULA FOR CONVERTING:

Number of Units × Conversion Number = New Number of Units

## TO CONVERT INCHES TO CENTIMETERS:

[number of inches] × 2.54 = [number of centimeters]

## TO CONVERT CENTIMETERS TO INCHES:

[number of centimeters] × 0.3937 = [number of inches]

## TO CONVERT FEET TO METERS:

[number of feet] × 0.3048 = [number of meters]

## TO CONVERT METERS TO FEET:

[number of meters] × 3.280 = [number of feet]

## TO CONVERT YARDS TO METERS:

[number of yards] × 0.9144 = [number of meters]

## TO CONVERT OUNCES TO GRAMS:

[number of ounces] × 28.35 = [number of grams]

## TO CONVERT GRAMS TO OUNCES:

[number of grams] × 0.035 = [number of ounces]

## TO CONVERT POUNDS TO KILOGRAMS:

[number of pounds] × 0.45 = [number of kilograms]

## TO CONVERT KILOGRAMS TO POUNDS:

[number of kilograms] × 2.2 = [number of pounds]

# INDEX

||||||||||||||||||||||||||||||||||||||||||||||||||||||||||||||||||||||||||||||||||||||||||||||||||